The Unconscious Helper Beings & Their Homes/Buildings

by
Chalvez Dickens

ISBN-978-1-960853-27-1

Liberation's Publishing LLC
West Point - Mississippi

This book is about Using Our Technology, (The Quantum Computer, The Mind Reading Machine, and The Replicator Machine all 3 Combined and Advanced enough) For Creating Buildings/Homes for the Unconscious Helper Beings to live and be made in. Basically the story is about and for us to get used to the technology that can create the homes/buildings for the Unconscious Helper Beings to live in, that the technology can create the buildings/homes in an Instance! Giving Freedom of All Beings in All Existences Forever with Old and New Freedoms Forever!

So it's no surprise when we actually achieve this goal in reality! All created for moral, ethical, comfortable usages...No one or no individual being should be put under any uncomfortable conditions or not even the smallest pain you can think of when using or creating the technology.

Architect: We need buildings and homes up in this city right now!

Quantum Computer

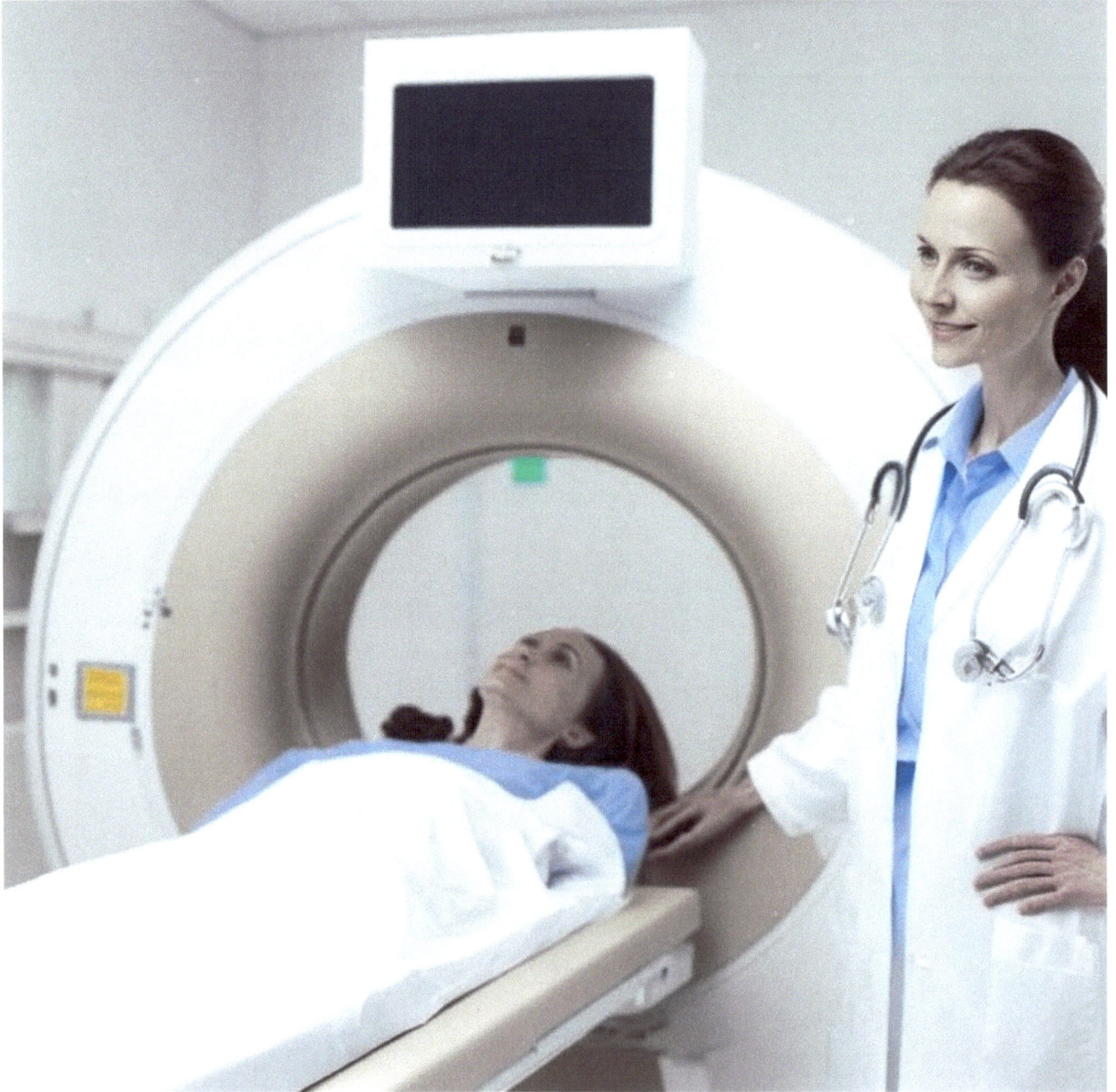

Mind Reading Machine

Replicator Machine

Student: Look the 3 Machines Combined! All created for moral, ethical, comfortable usages… No one or no individual being was put under any uncomfortable conditions or not even the smallest pain you can think of when using or creating the technology as well.

Architect: Ok perfect, tell the combined computer to create the Unconscious Helper Beings Buildings/ Homes with offices attached to it, so that they can be created in the offices and also live in the suites in the buildings, and neighborhoods!

Student: Computer…Create the Unconscious Helper Beings Buildings!

Narrator: In an instance the buildings and homes are up, giving freedom of all beings in all existences forever with old and new freedoms forever. Thanks to the Advanced Technology we have created, that allows buildings to be created in an instance, and can have to ability to create at the quantum scale, using mind reading technology, and the ability to create using the replicator. With all three combined, and advanced enough to understand all languages, you see there is nothing we can't accomplish. All we need now that we have the technology, is to input specific areas/locations into the computer so the computer knows where to place the buildings at the specific areas/locations.

Narrator: We call those designated areas....

Student: So anytime you need a new building or home, remember it can be built, in a instance, but the only thing you would need to input into the computer, is your designated area. That area must be ethically made appropriate for the objects and landmarks around it. It can be a small area, or it can be as large as All Existences.

Architect: Wow! The buildings and homes are up and finished with the Unconscious Helper Beings inside! They have families and eat, sleep, and think just like us!

Student: It should be anywhere around 1 trillion, 500 billion Unconscious Helper Beings buildings in the world, and that number is the same elsewhere but different compared to the population and areas the beings are located in other parts of existences, at around 81 meters tall. That number is for if only the buildings are being made, but they also can be in neighborhoods, there is a limit for a single neighborhood of Unconscious Helper Beings homes though for one being, so you might have to have a couple of neighborhoods if one choose to have it that way, only because we want our towns to not be as different as we normally see them every day. For the buildings though, they consist of 2,500 Unconscious Helper Beings per building.

That means it will be 150,000 buildings per 150,000 people, basically an entire designated area for a single being. Still 2,500 Unconscious Helper Beings per individual on Earth until we decide we want or need more, to make sure the planet is still in control of population and space for other buildings and creations. But remember the planet can and will grow if needed, peacefully. Also remember one can do anything they want with the land they are receiving, where the Unconscious Helper Beings buildings and homes would be, just in case one wouldn't want to use up all the land just for the Unconscious Helper beings buildings and their homes.

Student: The only difference between us and them is that their purpose is to work at the highest levels and have no want of free will and don't want to be free and are incapable of being free for forever. Also, they don't cost anything to receive them, and we get all the money from them working the jobs, and I know you are wondering how much money would that be. The answer is when you check your bank account, or wherever you put your money, you will see that it never ends. Not even the smartest being that ever existed can count it all. Their only motives when they wake up and go to sleep and when they dream, is to serve and work for all beings in all existences forever!

Architect: Wow! This is Amazing! We put those buildings and neighborhoods up fast!

Student: Thanks to the technology that allows us to put buildings up in an instance!

Architect: We have just accomplished a major milestone in history! The architect and student slap fives and says let's catch up later.

The buildings and homes in the city all up and running with Unconscious Helper Beings living, working and enjoying   themselves.

People/beings arriving and receiving services from the Unconscious Helper Beings and enjoying themselves.

The student finally says to an Unconscious Helper Being, "Can you take me to a new home made just like I like it, and   entertain me for the day." The Unconscious Helper Being says, "I know just what you like"

Then the student gets in the backseat of some expensive luxury spacious truck and the unconscious helper being begins to turn the music up for the student, while driving the student to his destination, passing the student a drink from out of the fridge in the vehicle, taking the student to which will be one of his new many homes.

Shows the student arrive at his Mansion at night! The Unconscious Helper Being opening the door to the truck…. The House is lit!

Narrator: Thank you Jesus and The Higher Powers Forever for giving Freedom of All Beings in All Existences Forever with Old and New Freedoms Forever!

www.ingramcontent.com/pod-product-compliance
Lightning Source LLC
Chambersburg PA
CBHW050912210326
41597CB00002B/100

9 781960 853271